P9-BYO-138

BUILDING AMERICA

The Alaska Pipeline

Craig A. Doherty and Katherine M. Doherty

A BLACKBIRCH PRESS BOOK
WOODBRIDGE, CONNECTICUT

To the memory of our friend, Donna Campbell

Special Thanks

The publisher would like to thank Tracy L. Green, Bob Schoff, and Lee Schoen of the Alyeska Pipeline Service Company for their valuable suggestions and cooperation on this project.

Published by Blackbirch Press, Inc.
260 Amity Road
Woodbridge, CT 06525
web site: http://www.blackbirch.com
e-mail: staff@blackbirch.com

First Edition

Printed in the United States

10 9 8 7 6 5 4 3 2

Editorial Director: Bruce Glassman
Senior Editor: Deborah Kops
Editorial Assistants: Laura Norton, Kristina Knobelsdorff
Design and Production: Laura Patchkofsky, Calico Harington

Photo Credits

Cover and pages 42–43: Fred Hirschmann, courtesy of Alyeska Pipeline Service Company; page 8: Clark James Mishler, courtesy of Alyeska Pipeline Service Company; page 11: James P. Blair/National Geographic Image Collection; page 40: ©Al Grillo/SABA Press Photos. All other photos courtesy of Alyeska Pipeline Service Company.

Library of Congress Cataloging-in-Publication Data

Doherty, Craig A.
The Alaska pipeline / by Craig A. Doherty and Katherine M. Doherty.
p. cm.—(Building America)
ISBN 1-56711-115-7 (lib.bdg.)
1. Trans-Alaska Pipeline (Alaska)—History—Juvenile literature. I. Doherty, Katherine M. II. Title. III. Series: Doherty, Craig A. Building America.
TN879.5.D64 1998 96–43442
388.5'5'09798—dc20 CIP
AC

Table of Contents

Introduction

Alaska has always been seen as the frozen North, a frontier that was too cold, except for the foolhardy. When Secretary of State William Henry Seward arranged for the United States to purchase Alaska from Russia in 1867 for $7.2 million, he was ridiculed by the press. The papers referred to Alaska as "Seward's Folly" and "Seward's Icebox." Even the discovery of gold in this remote part of the country didn't change many people's feelings about it. In spite of it all, Alaska became the 49th state on January 3, 1959.

The new state's importance to the "Lower 48," or the "outside," as Alaskans refer to the rest of the country, changed dramatically in 1968. A large oil deposit was discovered at Prudhoe Bay, on Alaska's North Slope (the land between the Brooks Range of mountains and the Beaufort Sea, in the Arctic Ocean). It was estimated that 10 billion barrels of oil were below the frozen tundra of Prudhoe Bay. The development of the North Slope oil field would be a great benefit to the economy of the United States, which at that time, was finding itself more and more dependent on foreign supplies of petroleum products.

There was one major problem facing the oil companies that wanted to drill for oil at Prudhoe Bay: getting the crude, or unrefined oil, from their wells in the North Slope to their refineries in the Lower 48. An early attempt to take a ship north of Canada through the Arctic Ocean, which is frozen for most

Opposite: *Thousands of segments of pipe were welded together to make the Alaska Pipeline, which was completed in 1977.*

of the year, proved to be a poor solution. Ice broke a hole in the ship, despite massive reinforcing of its bow (front).

The oil would have to be transported from the wells to an ice-free port. It is roughly 800 miles from Prudhoe Bay south to the port of Valdez. The only way to move a large volume of oil over that distance would be by pipeline. The United States is crisscrossed by oil and gas pipelines, but no one had ever built one through the obstacles that Alaska presented. Frozen ground, rugged mountains, and raging rivers were just some of the natural barriers that faced the designers of the project. Even more serious were concerns about the pipeline's impact on the environ-

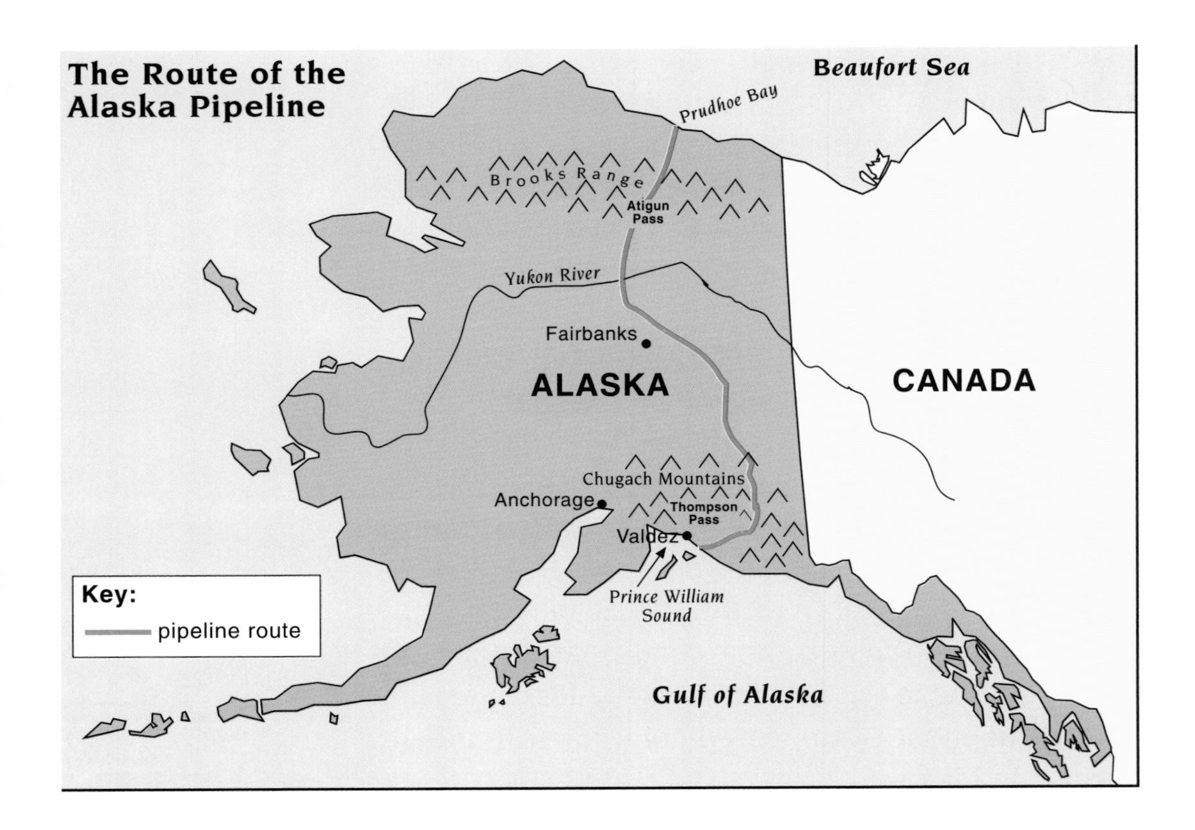

ment and the native Alaskans whose lands would be affected by the construction of the line.

In 1968, engineers and surveyors began laying out a possible route, and three oil companies—Atlantic Richfield, British Petroleum, and Humble Oil—formed the Trans Alaska Pipeline System (TAPS) Project. On February 10, 1969, in Anchorage, Alaska, the three companies publicly announced that they would be responsible for organizing, designing, and building the Trans Alaska Pipeline. As the plans began to take shape, it became clear that it would be the largest private construction project ever attempted. It would eventually cost over $8 billion. In 1970, a new company called the Alyeska Pipeline Service Company was created to execute the project. *Alyeska* is the ancient native word for Alaska, which means "the great land." The new organization brought more companies with leases to drill on the North Slope into the process of building the pipeline.

The pipeline crosses rugged mountains such as the Chugach range (top) *to reach the port of Valdez* (bottom).

The design phase of the project progressed rapidly, and the U.S. Department of the Interior appeared to be a strong supporter of the proposed plan. There were, however, others who felt the pipeline was wrong. Three environmental groups (the Wilderness Society, Friends of the Earth, and the Environmental Defense Fund) and five native Alaskan villages joined together to prevent construction. This conflict, and many others, would make the building of the Alaska Pipeline one of the most complicated construction undertakings in history.

CHAPTER

1

The Fight Begins

For much of its proposed route, the Alaska Pipeline would pass through land that was claimed by native Alaskan groups, such as the Iñupiat-Eskimos and Athabascans. Before construction could begin, there would have to be a settlement between the native Alaskans and the federal government in Washington, D.C. Native groups in Alaska had been treated poorly by the U.S. government in the past. Russia received $7.2 million in 1867 for land that really belonged to the native peoples of the North. The Alaskans were never adequately compensated for the loss of their lands, and they dreamed of a settlement even before oil was discovered in Alaska.

Opposite: *Environmental groups were very concerned that the pipeline would upset the delicate ecosystems of Alaska's unspoiled land.*

The oil lobby—people who worked in Washington, D.C., for the oil companies and represented their concerns and needs to the federal government—hoped the natives' claims would be settled as quickly as possible so that the oil companies could proceed with the pipeline. The oil companies wanted two things: the oil under the North Slope, and right-of-way, or legal right to build their pipeline across a strip of land running across the length of Alaska, from Prudhoe Bay to Valdez.

Settling the Claims

The Alaska Native Claims Settlement Act of 1971 provided the native people of Alaska with almost $1 billion and title to 44 million acres of Alaskan territory. The settlement also included rights to a small percentage of the oil companies' earnings from the pipeline. The bill passed through Congress easily on December 14, 1971, and was signed by President Richard Nixon on December 18. Many of the native Alaskans felt the deal offered to them was not in the best interests of all those involved.

The Iñupiat-Eskimos of the North Slope were the most dissatisfied with the provisions of the settlement act. They were upset by the fact that they would not get any of the land known to have oil reserves under it. Much of the Prudhoe Bay oil field was under their historic lands. Despite the Eskimos' opposition, the oil companies and their supporters in the federal government's Department of the Interior had cleared the first hurdle in their effort to get oil from the North Slope to the American consumer.

Protect the Environment?

The oil companies next had to deal with concerns expressed by the environmental groups. These groups worried that the pipeline would create ecological problems. Beneath the surface of much of the land in Alaska is a layer of earth called permafrost that is frozen year-round. The environmental groups were

The Eskimos attending this church service live on Barter Island, about 100 miles from Prudhoe Bay.

The Caribou Dilemma

One of the major concerns of environmentalists was the impact the pipeline would have on the rich and varied wildlife of Alaska. Bears, moose, sheep, wolves, caribou, and many smaller animals would be affected by the project. Studies were done, and much speculation was put forth on how the animals would react to the construction, and then, to the pipeline. As data was gathered, it became apparent that the greatest problem facing the line's designers was the caribou.

Caribou are the North American relatives of the reindeer, and they migrate in vast herds between their summer calving grounds—where the young are born—and their winter homes. There are 12 different herds of caribou in Alaska. Four range in the area of the pipeline: the Arctic, Porcupine, Delta, and Nelchina caribou.

The Arctic and the Porcupine herds' territories reached Prudhoe Bay, and the boundary between the two pretty much followed the proposed route of the pipeline. They could migrate to and from their calving grounds without crossing the pipeline.

The line presented a problem, however, for the few thousand Delta caribou and more than ten thousand Nelchina caribou. Both herds' migration routes crossed the pipeline. Wildlife experts speculated that the animals would not walk under the raised line. They thought even a calf separated from its mother would refuse to cross under the human-made structure.

Burying the pipe would create another set of problems. The designers realized that in areas where the permafrost layer was delicate, the heat from the oil flowing through a buried section of the line would thaw the permafrost.

The solution that finally won acceptance proposed running the pipeline above ground where the permafrost was unstable and then building 553 elevated crossings for the caribou. The crossings were at least 10 feet above the ground—4 feet higher than the average height of the raised pipeline. An additional 23 crossings were created along important caribou migration routes. These crossings had a minimum

afraid the heat from the hot oil in the pipe would melt the permafrost, softening the earth along the route of the pipeline. There was also concern that it would disrupt the migration patterns of the caribou that lived along the route. Environmentalists worried most about what effect oil spills from shipping accidents would have on ocean life.

width of 60 feet and were called "sag bends" because of the way the pipe dipped down to go underground. To build the crossings, the pipe was insulated and then buried. The native soils were removed and replaced with gravel that would stay firm, even if it thawed. Where it was impractical to replace the native soil with gravel, refrigeration plants were built, and chilled brine, or heavily salted water, was pumped through the ground in pipes buried under the crossing. The brine, which was chilled below freezing, kept the soil from thawing and allowed the animals to use the crossing.

The caribou are comfortable walking under the pipeline—much to the surprise of wildlife experts.

The environmentalists wanted to put a hold on the pipeline, or any other major project like it, until the state of Alaska completed a comprehensive land-use plan. Such a plan would take about five years to complete.

In February 1971, the Department of the Interior held hearings in Washington that lasted for three straight days. Over 100 witnesses, including representatives from all sides, and a number of government officials, testified. A week later, hearings were held in Anchorage, Alaska. Some of the testimony given in Anchorage and in Washington was critical of the construction project and the impact it would have on the land and animals of Alaska. The Department of the Interior ended up with 6,000 pages of testimony and other documents, plus over 1,000 pages of letters.

These Dall sheep were among the many wild animals that lived near the proposed pipeline.

The state of Alaska, like the oil companies, wanted the pipeline built. The economic base of Alaska was not very strong, and the revenues generated by the pipeline would be a help in solving all the state's financial difficulties.

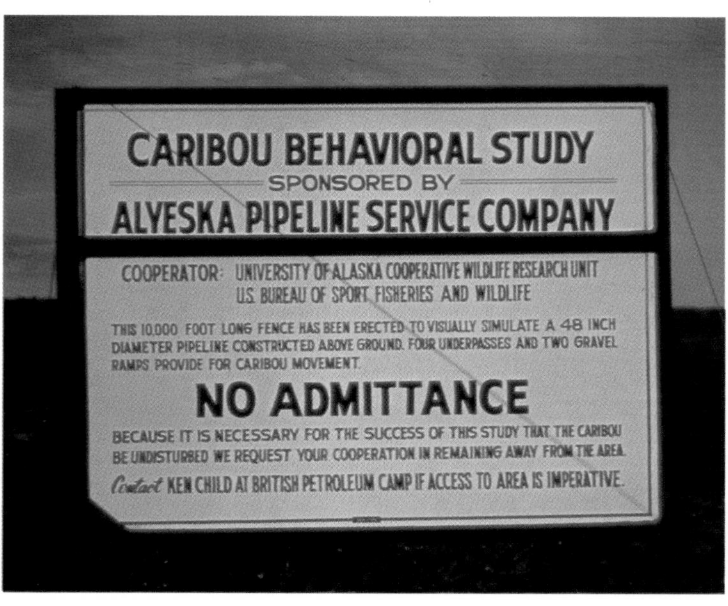

Wildlife experts studied the behavior of Alaskan caribou so that the pipeline's engineers could accommodate the animals' needs and habits.

The environmentalists did manage to halt construction of the pipeline, pending the release of the government's final report. On March 20, 1972, Secretary of the Interior Rogers C. B. Morton released the final environmental impact statement for the Trans Alaska Pipeline. The report suggested that the project would meet all the requirements of the National Environmental Protection Act (NEPA). Secretary Morton gave his final approval of the project on May 11, 1972.

The next round of the fight over the pipeline was battled in the courts. Conservationists tried to prove that the Department of the Interior did not have the legal authority to give Alyeska the right-of-way that the company needed to build the pipeline. On July 17, 1973, however, the Senate passed the Federal Land Rights-of-Way Bill allowing Alyeska the land it needed. The House did the same, and on November 16, 1973, President Nixon signed the bill into law. Construction of the pipeline could now begin.

CHAPTER

2

A Design Challenge

Even though Alyeska did not receive the go-ahead to start construction until November 1973, design and testing for the pipeline began almost immediately after oil was discovered in 1968. The engineers involved with the project had to deal with a variety of concerns. The biggest problems were related to the unique climatic conditions of Arctic Alaska. First among these was permafrost.

Opposite: *A repeating zigzag pattern enables the pipeline to accommodate the changing temperatures of the oil.*

In parts of the far North called the "tundra," the ground is permanently frozen, and the permafrost may reach a depth of 2,000 feet. On top of the permafrost lies about 18 inches of land, called the "active layer." In this top layer live the lichen and other plants that feed animals such as the caribou. It is a very fragile ecosystem; any disruption of the active layer of the tundra usually takes a long time to heal. If large areas of this layer are damaged, the permafrost underneath can thaw, especially in the summer. The engineers had to take the delicate nature of this soil in the North into consideration in order to avoid creating a bog across the length of Alaska.

One of the first decisions the engineers had to make was whether to design a cold or a hot pipeline. Oil comes out of the well at approximately 160 degrees Fahrenheit. The friction caused by oil flowing through the pipe and the pumping stations keeps it hot. Some lines chill the oil before it enters the pipe, and then again at various points along the way.

Designing a way for the pipeline to cross rivers such as the Gulkana was one of the many challenges that faced Alyeska's engineers.

"Bubbling Crude," "Black Gold," "Texas Tea" . . . Oil!

Oil, or petroleum as it is also called, plays an important part in the life of modern society. It runs our vehicles and generates electricity. Clothes, rugs, and other fabrics are made with it. Plastics are made from oil, and are a part of just about everything in our lives, from compact discs (CDs) to fighter jets. When it was first discovered, it was hoped that oil could be refined into a fuel that would be a less expensive substitute for the costly whale oil used in lamps. In 1852, a Canadian named Abraham Gessner patented a process for producing kerosene from crude oil. Ever since, people the world over have been searching for oil deposits.

Scientists believe that oil is formed at the bottom of oceans through a very long, natural process. As marine creatures of all shapes and sizes die, their remains pile up on the ocean floor and are eventually covered with sediment. As these layers pile up, the pressure on the bottom layers increases. The skeletons of the marine organisms become limestone, and the organic matter that made up the sea creatures is pressed into tiny globules, or balls, of oil. This process can take as long as 100 million years, and it is still going on today. A by-product of this process is a gas, called natural gas, which is also used as a fuel.

Most oil is trapped deep below a layer of shale or some other rock, and large deposits of crude oil lie waiting to be discovered. The Prudhoe Bay oil reserves—13 billion barrels of oil—lie between 8,000 and 10,000 feet below ground.

Workers drill through rock to create a new oil well.

Once the crude oil is pumped from the ground, it is transported by pipelines and huge ships, called supertankers, to refineries around the world. At these refineries, it is broken down into various forms that then can be turned into the thousands of products that we use every day. It is estimated that the world has only enough known oil reserves to get us about halfway through the next century. Because of this, people are constantly looking for undiscovered oil as well as alternatives to oil-based products and energy sources.

It was decided that a cold line in Alaska would be very difficult to maintain. Wax exists in all crude oil and builds up on the inside of cold lines, requiring them to be cleaned frequently. The engineers estimated that it would cost almost two thirds more to operate a cold line, so a hot line would be best.

A hot pipeline created its own unique set of problems. First among these was that an 800-mile-long heating element buried underground might thaw the unstable permafrost. It became apparent that the only feasible solution was to build an elevated hot oil line across areas of sensitive permafrost—which are present for about half the length of the pipeline.

Designing the support system for the raised sections of the line was the most difficult engineering task of the project. There were several problems: First, the pipe itself would expand and contract slightly as the temperature of the oil changed. More serious was the fact that the pipe would expand by as much as a mile when hot oil was first introduced into the newly built pipeline. This expansion could put enough pressure on the pipes to cause breaks. There was also a real fear that the vertical support members (VSMs) would absorb heat from the sun and transmit it to the permafrost.

To solve the expansion problem, the raised pipe was designed by the engineers to follow a repeating zigzag pattern in the shape of a trapezoid. The design relieved some of the stress caused by hot oil entering the new pipeline.

Where the VSMs came in contact with the permafrost, there needed to be some way to keep them from transferring the heat of the sun to the frozen

Most of the vertical support members were equipped with special pipes to protect the layer of permafrost below.

ground. The solution was to keep the ground frozen by super cooling it with special pipes in the footings of the VSMs. The pipes were filled with a special ammonia refrigerant, a waterless gas that cooled the ground. As the design phase continued, it was determined that 80 percent of the VSMs would require these pipes.

Thick layers of insulation were wrapped around the pipe to protect the oil from extremely cold temperatures.

The pipe that carried the oil was to be 48 inches in diameter, with 4 inches of insulation wrapped around it. The insulation was needed to keep the oil from cooling, especially if the line was ever shut down. The engineers determined that the insulation would keep the oil warm enough to withstand a 21-day shutdown at 20 degrees below 0 Fahrenheit.

Engineering most of the buried sections of the pipeline was straightforward and conventional. The top of the pipe was buried 3 to 35 feet below the surface. A bed of gravel was laid in the ditch before the pipe was set. The buried sections did not require the insulation that the above-ground line needed, except at certain caribou crossings. Instead it was coated with epoxy and then wrapped with plastic tape to prevent the pipe from corroding.

Selecting a Route

Engineers and surveyors fanned out over the Arctic wilderness in an attempt to determine the best route for the project. They decided that the pipeline should head south from Prudhoe Bay to the nearest ice-free port in southern Alaska. As the possible routes were analyzed, it became apparent to the engineers that using Valdez, Alaska, as the southern end of the pipeline was the most feasible choice. Much of this small port had been wiped out by an earthquake and an ensuing tidal wave in 1964. To make the oil terminal safe from a similar natural disaster, the oil storage tanks were built on a series of terraces dug into the side of the Chugach Mountains, just above the port.

CHAPTER

3

Building at Last

The specially made pipe for the line was delivered, beginning in 1969. It took over two years and 120 shiploads to get all the pipe to Valdez, Prudhoe Bay, and Fairbanks. Each ship carried approximately 8 miles of pipe in 40- and 60-foot sections. There were over 83,000 sections of 40-foot pipe. Half of these were sent to welders in Valdez, and the rest were sent to Fairbanks. In the relative comfort of their shops, the welders began joining sections together. Two 40-foot sections were joined to create 80-foot sections that would be welded together to make the pipeline. The pipe lay in storage until Alyeska got the go-ahead from Washington, D.C., to begin construction.

Opposite: *Workers joined long sections of pipe together before burying them in a ditch.*

When they first arrived at their port, these long lengths of pipe were shipped by rail to a welder's shop.

Work Begins

The first step in the process of actually building the pipeline was the construction of the haul road, which would enable workers to move heavy equipment where it was needed. This was to be the first road in the United States to cross the Arctic Circle. When construction finally started in 1974, the entire 358 miles of the haul road, from Prudhoe Bay to the Yukon River, were completed. It took 3,400 workers just over 150 days and a collective 3 million hours to complete the roadway. In addition to the labor involved, they used 32 million cubic yards of rock and gravel to create this passage in the middle of the wilderness.

The road was built from eight different starting points. Each of the crews kept track of their daily progress and compared their performance with the other crews along the line. Keeping track created a healthy sense of competition among the workers.

Saws that cut up the rock-solid frozen ground regularly needed to have their teeth replaced.

Helicopters and airplanes brought more than 160,000 *tons of materials and supplies for the construction of the pipeline.*

Workers built a 5-foot-thick ice bridge over the Yukon River so that they could drive across with supplies.

Before a permanent bridge was built over the Yukon River, the workers needed a way to get supplies to camps on the other side during the winter. To accomplish this, they built a 5-foot-thick bridge out of the most readily available material: ice. The river was already frozen, but the ice was not thick enough to support a truck. The engineers had the snow scraped off a 75-foot-wide strip of ice running from one side of the river to the other. Then they began spraying water from below the ice onto the strip. This had to be done when the air was cold enough for the water to freeze on contact with the existing ice. The ice was reinforced with logs and steel screens until it was over 5 feet thick and strong enough to support the largest equipment on the job.

X-rays and Welds

A critical aspect of the entire pipeline project was welding the pipes together. Welding is a way of joining two pieces of metal. A worker melts another piece of metal, called a welding rod, with a gas torch or electricity. The melted rod metal flows into the seam between the two pieces of metal being joined. When the heated metal cools, it forms a solid bond.

Poorly done welds are always a concern on a construction project because they are the weakest link in the structure. On the pipeline, bad welds could rupture and spill oil out on the ground, polluting the environment. Damaged pipe would also stop production because the line would have to be shut down until the pipe was repaired.

It was decided that every one of the almost 100,000 welds along the line would be inspected and X-rayed. Two contractors were hired to do the inspections. It turned out that one of them did not complete the work according to the company's contract with Alyeska.

In order to keep up with demanding schedules, the contractor took shortcuts and may have changed reports and films to make the results look better. Everyone from the president of the United States, Gerald Ford, down to the welders on the line got involved in this problem. President Ford sent a team to investigate, Congress held hearings, and Alyeska had to re-evaluate all the X-rays taken up until that time. Nearly 4,000 welds turned out to be questionable and had to be re-inspected and redone where necessary. Of all the questionable welds, only three were never

Laying the Pipe

On March 27, 1975, the first 1,900 feet of pipe was welded together and buried, and the actual pipeline was on its way. It had taken almost seven years from the time oil was discovered at Prudhoe Bay to get to this point, and it would take another two-and-a-half years before a tanker would leave the port of Valdez loaded with oil from the North Slope. Although there were many hardships and obstacles to overcome in constructing the line, the immediate task was to weld a seemingly endless number of pipe segments together.

Some welding jobs were done right inside the pipe.

thoroughly rechecked. These were 17 feet below the Koyukuk River, just south of the Brooks Range. The welding repair and re-evaluation program added an estimated $55 million to the cost of the project.

The largest obstacles were the mountain passes that the pipeline had to cross. One of the most difficult passes was Keystone Canyon, north of Valdez. In just 4 miles, the line had to climb up 800 feet and then drop down another 800 feet. The 80-foot sections of pipe were too heavy to lift with a helicopter, and the regular trucks could not climb the steep and twisting access road that ran alongside the pipeline. Workers had to build a special device for moving the pipe up the canyon. They attached cradles for the pipe to the sides of a bulldozer. Then a second bulldozer was attached to the one carrying the load

A cable tramway was built to get workers and equipment up the steep Thompson Pass.

to help push it up the steep slope. It took the two bulldozers about half an hour to get a load up to the crews, and it took them about a week to move all the pipe into place.

As the crews made progress on the pipe, other crews built the pumping stations that would keep the oil moving. There are nine stations pumping oil along the line, and each one is a maze of pipes, valves, tanks, and pumps. Four stations move the oil from Prudhoe Bay up to the top of the Brooks Range at Atigun Pass. Four move the oil south past Fairbanks and over the Alaska Range. There, another station pushes the oil up to the top of Thompson Pass, and gravity pulls the oil down to Valdez.

After working an average of 10 hours or more, most workers returned "home" to their cold, isolated construction camps.

The finished pipeline gleams under a breath-taking Alaskan sunset.

The pumping stations in the permafrost areas were built on insulated pads. Each of these stations has its own refrigeration plant to pipe chilled brine through the frozen ground. This prevents the ground from thawing. In addition to the huge pumps that move the oil, each station has a 55,000-barrel storage tank so that pressure can be easily released from the line in the event of an emergency. Each station generates its own power to run the pumps, but all the stations are linked together by communications equipment to the main control station at Valdez.

Testing the Line

As sections of the pipeline were completed, they needed to be tested to be sure that they could withstand the pressure of the oil being pumped through them. To do this, a section of pipe was sealed off from the rest of the line, and water was pumped in until the pressure inside the pipe reached 125 percent of the maximum pressure at which the oil would be pumped. The working pressure inside the line would reach up to 1,180 pounds per square inch (psi).

As the end of the 1977 construction season approached, the pipeline was almost finished. Alyeska and its contractors put up for sale most of the $800 million worth of equipment they had purchased, including cranes, backhoes, and bulldozers. The camps along the line began to close down once the testing of a section was successfully completed. The final stages of the construction involved rewelding a number of joints and a general cleanup of the entire 800 miles of the construction route.

CHAPTER

4

Turning On the Pumps

As the pipeline neared completion, another project was furiously under way at Prudhoe Bay. Hundreds of miles of additional pipe had to be set up to get the oil from the wells to the main line. Almost all oil wells produce natural gas as well as crude. Some of the gas coming out of the Prudhoe Bay wells was used to fuel power plants at the oil field. A gas line was also built parallel to the first 148 miles of the pipeline to provide fuel for four of the pumping stations. As the final preparations were completed along the pipeline and in the oil field, Alyeska prepared to open the valves and begin sending oil to Valdez and the waiting tankers.

Opposite: *In June 1977, the press, visitors, and workers gathered to feel the first oil coming through the line.*

A special device called a "pig" helped to get the air out of the pipeline before it was filled with oil.

Oil in a pipeline cannot be turned on like water in a faucet. Starting the oil flowing through the pipe requires a slow and careful approach. One method of starting up a pipeline is to first fill the line with water. Then, once all the air is out of the line, a device called a pig is placed in the pipe before the oil is pumped. A pig fits snugly into the pipe and travels down the line in front of the oil. There are even special pigs used to keep the inside of the pipe clean. All the air has to be removed from the line before the oil can be pumped because the oxygen in the air can react with the oil and cause it to become explosive.

The pipeline engineers had to come up with a new way to get the air out of the pipeline because they were afraid if they used water, it might freeze, and the line would become one big ice cube. Their solution was to fill the first section of the line with nitrogen and then slowly replace the nitrogen with oil. As the nitrogen was released from one section, it would fill the next section.

The first oil went into the main line on June 20, 1977, and was pumped forward very slowly. The pig, which was between the oil and the nitrogen gas, moved forward at a speed of 1 mile an hour. An inspection crew walked along the pipeline, following its progress. It was their job to watch for leaks and other problems. At this slow rate, it would be more than a month before any oil would reach Valdez.

The first oil arrived in Valdez at the end of July. It then took another 120 days to get the oil up to the correct temperature and moving at the right speed. At peak production, 2.1 million barrels of oil a day reached Valdez and its waiting tankers. The first tanker to leave Valdez with a full load of oil from Prudhoe Bay was the ARCO *Juneau*, which sailed on August 1, 1977.

Top: *Representatives of Alyeska and the Valdez Chamber of Commerce stand with the first barrel of crude oil.*
Bottom: *The ARCO* Juneau *left Prudhoe Bay with the first load of oil on August 1, 1977.*

The Exxon Valdez Oil Spill

Everyone's biggest fear about the Alaska Pipeline came true on March 24, 1989, when the *Exxon Valdez* ran aground in Prince William Sound and spilled 260,000 barrels of oil into the water. Over 10 million gallons of thick crude oil polluted the waters and shores of the sound.

The oil spill spread rapidly, and the response to the disaster was slow. It took emergency crews over 14 hours to arrive at the ship. It was another 7 hours before floating oil containment booms were in place. (These floating tubes keep most of the oil from floating away from the spill site.)

Over 100,000 geese, ducks, and other birds were killed by the Valdez oil spill.

Then a violent storm in Prince William Sound slowed the work of the crews. By the time it was over, the worst oil spill in U.S. history had begun killing many of the creatures in its path. Within three weeks, the oil from the *Exxon Valdez* had reached the Katmai National Park, over 250 miles away, and eventually affected 1,100 miles of Alaska's coastline.

It is estimated that over 100,000 birds were killed by the oil. About 6,400 otters were known to have died, which was about half the otter population in Prince William Sound.

There were many inquiries into the disaster. The captain of the *Exxon Valdez*, Joseph J. Hazelwood, had apparently turned into the inbound channel to avoid ice in the water. Then he turned on his autopilot, which steers a boat automatically, and accidentally ran his ship onto some submerged rocks. The rocks ripped holes in 10 of the ship's 15 oil tanks.

It was later revealed that Captain Hazelwood had been drinking that evening. His negligence and his employer's failure to enforce maritime laws resulted in the terrible accident that made the environment of Prince William Sound a disaster zone. Exxon was forced to pay over $1 billion in fines and damages to help with the cost of the cleanup.

Keeping the Oil Flowing

The 1.5 million barrels of oil flowing through the Alaska Pipeline each day generate approximately $2.5 billion in revenues a year. They also generate a number of operational problems. The procedures for checking the line have come into question a number of times. In 1993, the situation had become so serious that Congressional hearings were convened in Washington, D.C., to investigate the effectiveness of the safety-and-checking procedures for the line.

The hearings determined that substantial improvements would need to be made in the way that the pipeline was inspected and maintained. Alyeska has since made an increased effort to address these problems.

In addition to Alyeska's maintenance problems, 9 miles of pipe corroded. Although the pipe was wrapped in heavy-duty tape, some water in the soil seeped through to the metal. These sections of buried pipe were replaced. Alyeska continues to monitor the system with corrosion-detection pigs and to replace corroded pipe when necessary. Most of the pipes that collect toxic vapors from the tops of the tanks at the Valdez terminal will also be replaced. This piping system is at the end of its natural life span.

The production of oil on the North Slope will decline in the future. By the year 2000, Alyeska will operate with six of the pumping stations along the line. As the oil at Prudhoe Bay is used up, geologists and other scientists will continue to search for as-yet-undiscovered sources of oil, as well as alternative sources of energy.

Winding its way through Alaska's rugged wilderness, the pipeline still carries millions of barrels of oil through its veins each day.

Glossary

ammonia refrigerant A waterless gas used to cool other substances.

bog A wet, swampy area.

brine Water to which a large quantity of salt has been added.

cable tramway A moving, overhead cable to which items can be attached and then lifted up a slope.

caribou A large, antlered mammal found in the northern parts of North America. It is related to the reindeer.

crude oil Oil as it comes out of the ground, before it is refined.

ecological disaster A natural or human-made problem that has a severe, negative impact on the natural world.

ecosystem An environmental community that includes all the plants, animals, and natural features of an area that function together as a unit.

haul road The gravel road built from the Yukon River to Prudhoe Bay to allow construction of the Alaska Pipeline.

insulation A material used to prevent the transfer of heat or cold from one object to another.

oil There are many kinds of oils. In this book "oil" is petroleum, a flammable liquid refined into fuel to run cars and other equipment. It is also used to make a wide variety of products, including plastics.

permafrost A permanently frozen layer found below the earth's surface in some locations.

pig A device that fits inside a pipeline and can be used to separate different substances in the line, to clean the inside of the pipe, or inspect it for corrosion.

pipeline A line of welded pipe with valves and pumps that is used to transport liquids or gases.

pumping station A structure that contains the machinery to keep the oil flowing along the pipeline.

right-of-way Legal rights to a strip of land.

terminal A large storage facility located at the end of the pipeline.

trapezoid A four-sided shape with only two sides parallel.

U.S. Department of the Interior The executive branch of the federal government that oversees the use of land in the United States.

vertical support member (VSM) A structure built to support elevated sections of the pipeline.

welder A worker who is expert at welding, or using heat to join pieces of metal.

CHRONOLOGY

1968 Discovery of oil at Prudhoe Bay announced.

1969 February 10 Construction of Alaska Pipeline announced.

April Pipe is ordered.

June 1 Right-of-way for pipeline application submitted.

1970 Alyeska Pipeline Service Company is formed.

Five native Alaskan villages sue to stop pipeline; three environmental organizations join suit.

1971 February Hearings on pipeline are held.

September 13 First pipe for pipeline is delivered.

December 18 Alaska Native Claims Settlement Act of 1971 signed by President Richard M. Nixon.

1972 March 20 U.S. Department of the Interior issues final environmental impact statement.

May 11 Secretary of the Interior Rogers C. B. Morton approves construction of the pipeline.

1973 November 16 Federal Rights-of-Way Bill signed by President Nixon, clearing the way for pipeline permit approval.

1974 The haul road is completed. Construction begins.

1975 March 27 First section of pipe is buried.

1977 June 20 First oil starts to flow in pipeline.

August 1 ARCO *Juneau* is first tanker to depart Valdez with oil.

1989 March 24 *Exxon Valdez* oil spill in Prince William Sound near Valdez.

Further Reading

Alexander, Bryan and Cherry Alexander. *Inuit*. Chatham, NJ: Raintree Steck-Vaughn, 1992.

Bailey, Donna. *Energy from Oil and Gas*. Chatham, NJ: Raintree Steck-Vaughn, 1990.

Fradin, Dennis B. *Alaska—From Sea to Shining Sea*. Danbury, CT: Children's Press, 1993.

Harris, Lorle K. *The Caribou*. Morristown, NJ: Silver Burdett Press, 1989.

Heinrichs, Ann. *Alaska*. Danbury, CT: Children's Press, 1990.

Lampton, Christopher. *Oil Spill*. Brookfield, CT: Millbrook Press, 1992.

Schouweiler, Thomas. *The Exxon-Valdez Oil Spill*. San Diego, CA: Lucent Books, 1991.

Younkin, Paula. *Indians of the Arctic and Subarctic*. New York: Facts On File, 1991.

Source Notes

Allen, Lawrence, J. *The Trans Alaska Pipeline*. Seattle: Scribe Publishing Corporation, 1975.

"Alyeska Has Fallen Behind on TAPS Remedial Efforts." *Platt's Oilgram News*, February 26, 1996, 2.

"Alyeska to Replace Lines." *Oil Daily*, May 20, 1996, 5.

"Alyeska, TAPS Operator, Plans the First Closure of a Pump Station." *Oil and Gas Journal*, December 5, 1994, 2.

Berry, Mary Clay. *The Alaska Pipeline: The Politics of Oil and Native Land*. Bloomington, IN: Indiana University Press, 1975.

Bowermaster, David. "A Long, Blotted Record." *U.S. News & World Report*, October 25, 1993, 38–39.

Cullison, Andrew. "Corrosion Vigilance Keeps Alaskan Oil Flowing." *Welding Journal*, May 1991, 27–32.

"Database." *U.S. News & World Report*, January 15, 1996, 14.

Dixon, Mim. *What Happened to Fairbanks? The Effects of the Trans-Alaska Oil Pipeline on the Community of Fairbanks, Alaska*. Boulder, CO: Westview Press, 1978.

Egan, Timothy. "Alaska Pipeline faces costly repairs." *New York Times*, March 11, 1990, 22.

Halevi, Marcus, photographs and text by Kenneth Andrasko. *Alaska Crude: Visions of the Last Frontier*. Boston: Little, Brown, 1977.

Horton, Tom. "Paradise Lost." *Rolling Stone*, December 14, 1989, 150–182, 241.

"Integrity of Alaska Pipeline Questioned." *Science News*, December 4, 1993, 382.
Lieberman, Jethro K. *Checks and Balances: The Alaska Pipeline Case.* New York: Lothrop, Lee & Shepard, 1981.
McGrath, E. *Inside the Alaska Pipeline.* Millbrae, CA: Celestial Arts, 1977.
Mead, Robert Douglas. *Journeys Down the Line: Building the Trans-Alaska Pipeline.* Garden City, NY: Doubleday, 1978.
Miller, William H. "TAPS in Retrospect." *Industry Week*, April 29, 1985, 32.
"Pipeline Is Corroded." *National Geographic*, August 1990, p. preceding p. 1.
"Regulators OK Alyeska Plan." *Oil Daily*, April 11, 1996, 5.
Roscow, James P. *800 Miles to Valdez: The Building of the Alaska Pipeline.* Englewood Cliffs, NJ: Prentice-Hall, 1977.
Skorupa, Joe. "Driving the Alaska Pipeline Road." *Popular Mechanics*, October 1995, 74–77, 119.
"Workers Excavate Pipeline." *Oil Daily*, April 24, 1996, 5.

INDEX